Insects
and
Spiders

Dragonflies

Shane F McEvey
for the Australian Museum

This edition first published in 2002 in the United States of America by Chelsea House Publishers, a subsidiary of Haights Cross Communications.

Chelsea House Publishers
1974 Sproul Road, Suite 400
Broomall, PA 19008-0914

The Chelsea House world wide web address is www.chelseahouse.com

Library of Congress Cataloging-in-Publication Data Applied for.

ISBN 0-7910-6597-9

First published in 2001 by
Macmillan Education Australia Pty Ltd
627 Chapel Street, South Yarra, Australia, 3141

Edited by Anna Fern
Text design by Nina Sanadze
Cover design by Nina Sanadze
Australian Museum Publishing Unit: Jennifer Saunders and Catherine Lowe
Australian Museum Series Editor: Deborah White

Printed in China

Acknowledgements
Our thanks to Martyn Robinson, Max Moulds and Margaret Humphrey for helpful discussion and comments.

The author and the publisher are grateful to the following for permission to reproduce copyright material:

Cover: A resting dragonfly, photo by John Kleczkowski/Lochman Transparencies.

Australian Museum/Nature Focus, pp. 26, 27; Babette Scougal/Nature Focus, p. 10 (top); Densey Clyne/Mantis Wildlife, pp. 7 (top), 15 (bottom), 18 (top); Dominic Chaplin/Nature Focus, p. 28 (right); Hans & Judy Beste/Lochman Transparencies, p. 22 (left); Jim Frazier/Mantis Wildlife, p. 22 (left); Jiri Lochman/Lochman Transparencies, pp. 4, 13 (bottom), 14 (bottom), 29 (top), 30; John Kleczkowski/Lochman Transparencies, pp. 5 (top and bottom), 6–7, 12 (top and bottom), 16 (top and bottom), 18 (bottom), 24 (top), 28 (left); Kate Lowe/Nature Focus, p. 9 (top); Kathie Atkinson, pp. 8, 9 (bottom), 11 (top and bottom), 23 (top), 24 (bottom), 25 (bottom), 29 (bottom); Kathie Atkinson/Auscape, p. 19 (middle); Mark Newton/Auscape, p. 10 (bottom); Pascal Goetgheluck/Auscape, p. 21 (top); Pavel German/Nature Focus, p. 23 (bottom); Peter Cook/Auscape, p. 25 (top); Peter Marsack/Lochman Transparencies, p. 28 (middle); Raoul Slater/Lochman Transparencies, pp. 13 (top), 17 (right); Reg Morrison/Auscape, pp. 13 (middle), 19 (top); Steve Wilson, p. 20 (all); Steven David Miller/Nature Focus, pp. 14 (top), 15 (top); Stephen Richards/Nature Focus, pp. 17 (left), 19 (bottom); Wade Hughes/Lochman Transparencies, p. 21 (bottom).

Contents

Glossary words

When a word is printed in **bold** you can look up its meaning in the Glossary on page 31.

What are dragonflies?

Dragonflies are insects. Insects belong to a large group of animals called invertebrates. An invertebrate is an animal with no backbone. Instead of having bones, dragonflies have a hard skin around the outside of their bodies that protects their soft insides.

Dragonflies have:
- six legs
- four wings
- two **antennae**
- two eyes
- a mouth
- many breathing holes on the sides of their bodies.

A male dragonfly grabs hold of a twig with his strong legs.

What makes dragonflies and damselflies different from other insects?

Dragonflies and **damselflies** are large insects with two pairs of wings. The wings have a network of veins and a dark mark near the tip. They have a very long, slender **abdomen**. They are **predatory** insects. Their bristly legs point forward, ready to capture prey in mid flight.

They are usually found near water where their **nymphs** live. When the male and female mate, they hold each other with their abdomens looped in a special way.

Scientists have given a special name to all dragonflies and damselflies. They are called **Odonata**. The two kinds of Odonata, dragonflies and damselflies, can be easily recognized.

Dragonflies

- Dragonflies have different shaped front and back wings. These wings are always held straight out from their bodies.
- Dragonflies are strong fliers and have very large round eyes that are so large they almost meet at the top of their heads.
- Dragonfly nymphs have three tails.

A yellow dragonfly perches on a plant.

Damselflies

- Damselflies' front and back wings are similar in shape. Damselfly wings are often folded back over their long abdomens.
- Damselflies are weak fliers and they rest often.
- Their eyes are not as large as dragonfly eyes and are further apart on their heads.
- Damselfly nymphs have four tails.

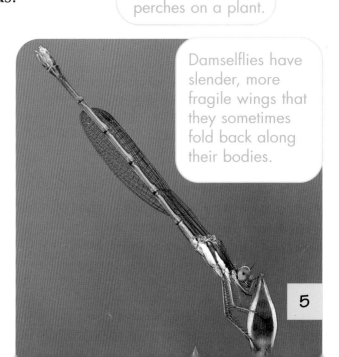

Damselflies have slender, more fragile wings that they sometimes fold back along their bodies.

5

Dragonfly bodies

The body of an adult dragonfly is divided into three segments. These segments are called the head, the **thorax** and the abdomen. The abdomen of an adult dragonfly is very long.

Adult dragonflies have short hairs on their bodies and around their mouths.

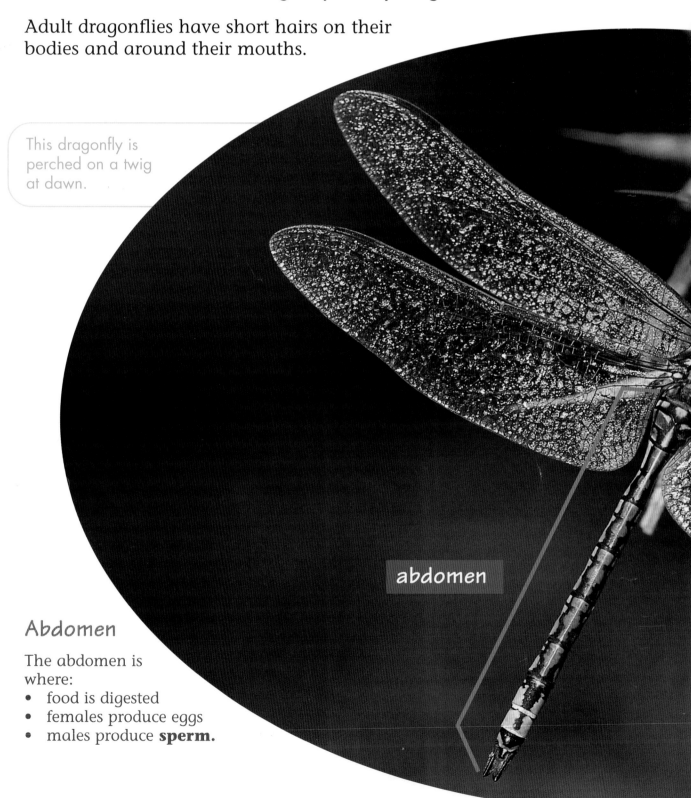

This dragonfly is perched on a twig at dawn.

abdomen

Abdomen

The abdomen is where:
- food is digested
- females produce eggs
- males produce **sperm.**

Dragonflies have some of the brightest colors in the insect world. This dragonfly is bright red.

head

thorax

Head

On the head are the:
- mouth
- **mandibles**
- antennae
- eyes.

Fascinating Fact

Dragonflies are the fastest flying insect. They can fly at over 30 kilometers (18 miles) per hour.

Thorax

On the thorax are:
- legs
- wings.

The head

On the head of an adult dragonfly are the mouth, eyes and antennae.

Mouth

Dragonflies have mandibles for biting and chewing.

Eyes

Dragonflies have two very large compound eyes that sometimes meet in the center. Damselflies have compound eyes that do not meet. A compound eye is made up of lots of tiny eyes packed together. Dragonflies can see all around themselves. They have very good eyesight. The large compound eyes of a dragonfly can be made up of many thousands of tiny eyes.

Antennae

Dragonflies have very small, simple antennae.

Dragonflies have hairs around their mouths. These hairs help them feel and possibly taste their food.

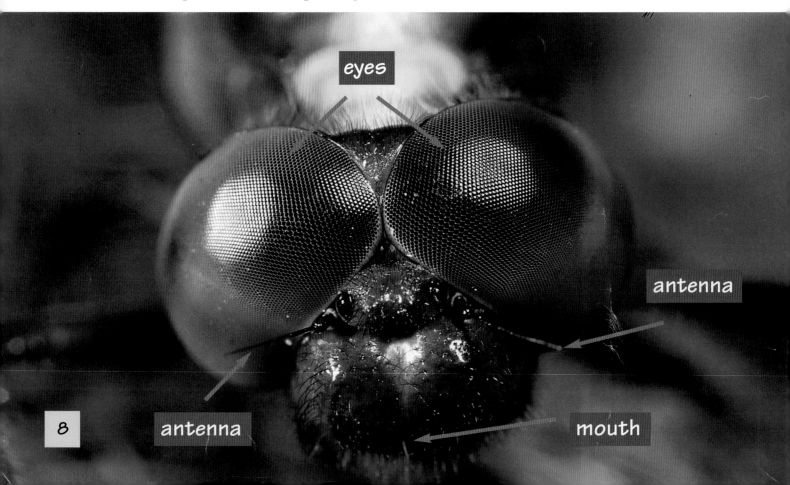

eyes

antenna

antenna

mouth

The thorax

On the thorax of an adult dragonfly are the legs, wings and some of the breathing holes.

Legs

Dragonflies have six legs. They use their legs for landing, perching and catching their prey. Adult dragonflies do not walk. All six of the dragonfly's legs form a basket-like trap that faces forward. This allows the dragonfly to fly straight at its prey and grab it in mid-flight.

legs

A dragonfly catches its prey by trapping it inside a cage it makes with its legs.

Wings

Dragonflies have two pairs of large wings. Each wing has a **membrane** with a network of hard veins. The veins support the wings so they can be used for flying. Sometimes the wings are clear and sometimes they are brightly colored.

Damselflies have front and back wings that are the same size and shape. Dragonflies have different shaped wings — the back ones are rounded near the base and the front ones are narrow near the base. Dragonflies have very powerful wings that allow them to fly very, very fast and over long distances. They can dart, glide and hover. Sometimes they can even fly backwards.

Breathing holes

Dragonflies breathe through tiny holes in the sides of their bodies called spiracles. Dragonflies do not breathe through their mouths.

front wings

back wings

The back wings of a dragonfly are rounded near where the wing attaches to the dragonfly's body.

Did you know?

A dragonfly's wings make a lot of noise when the dragonfly is caught or trapped. Although the wings look fragile, they do not break easily.

Where do dragonflies live and what do they eat?

Dragonflies and damselflies almost always live near fresh water. This could be a fast-flowing stream or river, a swamp, pond, dam or lake.

If the water dries up, the dragonfly is able to fly large distances to find another suitable water **habitat**. Dragonflies can even find pools of water in the middle of deserts.

Dragonflies live around bodies of water like ponds. They can often be seen perched on the reeds that grow around the pond.

This blue dragonfly is feeding on a fly that it has caught.

Hunting for food

Dragonflies and damselflies are hunting predators. They hunt and catch their prey while they are flying. Dragonflies and damselflies eat other flying insects like mosquitoes, **midges** and **caddisflies**. They find their food by seeing it. Of all the insects, dragonflies are the best at hunting and catching prey while they fly.

When a dragonfly or damselfly catches its prey, it forms a cage with its legs so the prey is trapped inside. If the prey struggles, the dragonfly can bite off the prey's legs and wings. The dragonfly can then take its time to eat the prey.

Fascinating Fact

Large dragonflies will sometimes catch and eat damselflies, which are smaller and less powerful fliers. Any flying insect is at risk when passing through a dragonfly habitat.

Dragonflies can often be seen perched on a vantage point, scanning the area around them for insect prey.

Did you know?

Even stinging bees and wasps can fall victim to a dragonfly.

This dragonfly has caught a fly.

11

Dragonflies that live in deserts and dry habitats

Dragonflies can live in very hot and dry places like deserts as long as there is suitable water. Here are some of the dragonflies that can live in hot, dry places.

These dragonflies can be found all over Australia. When it rains in inland Australia or when the inland rivers flood, dragonflies can take advantage of the vast wetland habitats.

Adult males of this kind of dragonfly are always red in color. These dragonflies can fly long distances in search of water. Occasionally they will rest on rocks or plants.

Dragonflies can live near fresh water wherever it occurs. In dry areas, they can make use of farm dams. This dragonfly is perching on a wire fence while it watches for prey.

Dragonfly eggs may survive buried in mud until rain and floods make the water level rise again.

This dragonfly is stuck in the mud of the Coongi Lakes in arid Central Australia.

The thorax of insects, especially strong fliers like dragonflies, are packed with muscles. These muscles make the wings work. There are separate muscles for the front wings and for the back wings. These strong muscles enable dragonflies to fly great distances in search of water.

Dragonflies that live in forests and wet habitats

Many dragonflies and damselflies like to live in forests. Here are some that live around water in forests.

Is this dragonfly drinking nectar from the flower? Probably not. The buzz of hundreds of other insects flying around the bottlebrush tree best explains why the dragonfly perches there. It is watching for a passing insect to catch and eat.

Tropical insects are sometimes beautifully colored, like this dragonfly from northern Australia. Different kinds of dragonflies and damselflies have different colors and patterns.

Did you know?

Some dragonflies and damselflies have clear wings without any color at all.

Some dragonflies look like the things they frequently sit on. This is one way they can avoid being seen by even larger predators like birds. This dragonfly is bright green like the leaves under it.

Damselflies have two pairs of wings that are similar in shape and size. This is what makes damselflies different from dragonflies. The color patterns on their bodies are also important in identifying what kind of damselfly it is. This damselfly is perched on leaf litter lying on the forest floor.

How dragonflies communicate and explore their world

Dragonflies get information about their environment mainly by sight. They find their prey and other dragonflies by seeing them.

Most dragonflies are active during the day when the sun warms their bodies but some are active at dawn and dusk. In the warm tropics, some even fly at night, but they cannot hunt in the dark.

In cooler parts of the country, dragonflies and damselflies are active only during the warmer months of the year. In the warm tropics, dragonflies and damselflies are active all year round.

In some dragonflies, the eyes are so large they actually touch at the top of the head. Damselfly eyes are never this big.

Dragonflies usually sit with their wings spread out like this. Damselflies usually fold their wings back along their bodies when they are resting. At dawn, when they are still cold and sometimes covered in dew, dragonflies and damselflies are not active. During the morning, their wings will dry and the sun's warmth will make them active.

How dragonflies communicate

Dragonflies rely on their vision to communicate with each other.

- The males see other males and defend their territory.
- The males see females and mate with them.

Male dragonflies and damselflies are **territorial.** This means that they will live in a particular space and fight off other males that enter. They do this by keeping watch, patrolling their chosen territory and attacking any intruders.

Did you know?

Dragonflies beat their wings about 25 times per second, which is much slower than flies, bees and wasps but faster than most moths and butterflies.

A group of dragonflies perch on a reed near a pond. Dragonflies sit with their bodies straight out because their legs go forward from their thorax, not down.

The life cycle of dragonflies

The whole life cycle of a dragonfly or a damselfly, from egg to adult, can take a few months or as long as several years.

Dragonflies reproduce **sexually**. This means that a male and a female are needed to make new dragonflies. The male dragonfly provides sperm while the female dragonfly provides eggs. The eggs and sperm need to join together for a new dragonfly to start growing.

Male dragonflies find female dragonflies by looking for them. Some male dragonflies are brightly colored and this probably attracts the attention of females.

A pair of dragonflies mating. The male has his abdomen out straight and the female is underneath with her abdomen curled up towards the male.

The new adult dragonfly will sometimes remain sitting on its nymphal skin until its wings harden, its abdomen expands and it can fly. Once they are adult, dragonflies do not grow any more.

Did you know?

Dragonfly nymphs can live for up to five years and are fun to keep in fish tanks at home.

A new adult dragonfly sitting on its empty nymphal skin.

18

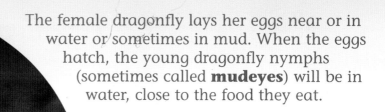

The female dragonfly lays her eggs near or in water or sometimes in mud. When the eggs hatch, the young dragonfly nymphs (sometimes called **mudeyes**) will be in water, close to the food they eat.

Dragonfly nymphs are shorter and fatter than adult dragonflies. They have longer antennae and no wings. They have smaller eyes and usually live in water. Damselfly nymphs are similar except they have four tails instead of three.

Female dragonflies sometimes lay many hundreds of eggs.

The nymphs spend most of their time eating and growing bigger. Nymphs are predators. They eat little water animals. As the nymphs grow, their skin becomes very tight until it splits. This allows the nymph to grow even bigger in a new, larger skin. This is called **molting** and it happens several times during the life of a nymph.

Dragonfly and damselfly nymphs are often found in freshwater ponds. They look very different to the adult dragonflies and damselflies.

When a nymph is ready to turn into an adult, it crawls out of the water and onto a plant. It then breaks out of its last nymphal skin and becomes an adult dragonfly.

This dragonfly has just broken free from its nymphal skin.

Remarkable mating behavior

One of the things that is special about dragonflies and damselflies is how they mate. When the male is ready to mate, he transfers sperm from the tip of his abdomen to a special place under the other end of his abdomen, near the thorax. He then grabs a female behind her head using special claspers on the end of his abdomen. The female reaches forward with her abdomen to get the sperm from the special place on the male.

Sometimes the male will keep holding the female until she lays her eggs. You often see pairs of dragonflies or damselflies flying around or sitting connected.

A male damselfly holds onto the leaf of a plant with his legs and, with special claspers on the end of his abdomen, he holds a female damselfly behind her head.

The female damselfly, while still being held by the male, curls her abdomen up until it touches the male damselfly's body where he has put some sperm.

Nymphs

Dragonfly nymphs, or mudeyes, live in water and move around by walking. They eat water animals like fly and beetle **larvae**, water bugs, tadpoles and even small fish. Mudeyes are often eaten by large fish. Anglers often use mudeyes for bait.

Mudeyes often catch their prey by sneaking up and grabbing it. All mudeyes have a special lower lip that they keep folded in front of their face. This lip has a cage of spines at the end. The mudeye can suddenly shoot this lip forward and grab its prey with the spines. It then pulls the prey back to its mouth.

Some dragonfly nymphs come out of the water and feed on small land animals. Some nymphs live their entire life out of the water.

This mudeye has its lower lip extended out. You can see the sharp spines at the end.

Fascinating Fact

Mudeyes eat mosquito larvae so they are good to have in ponds near where you live.

Breathing underwater

Mudeyes can breathe underwater. Some mudeyes have breathing organs outside their bodies. Others have a type of 'lung' inside their body.

Mudeyes with breathing organs outside their body have feather-like tails. These tails are used for breathing. They absorb oxygen from the water.

Mudeyes with breathing organs inside their bodies breathe by sucking water into their body through their **anus**. The water is expelled after the mudeye has absorbed the oxygen from it.

This dragonfly nymph has three tails on the end of its abdomen.

Predators and defenses

Dragonfly and damselfly nymphs are eaten by fish, frogs, birds and reptiles.

Adult dragonflies and damselflies are mostly eaten by large spiders, frogs, birds and bats.

Dragonflies are strong, fast fliers. They are very hard to catch. Damselflies are weak fliers and are sometimes eaten by dragonflies.

Dragonflies are strong insects. It takes a big spider to be able to catch them. This dragonfly has been trapped in the web of a large orb-weaving spider.

Sometimes damselflies are caught by birds. This bird has caught a damselfly to feed to its hungry chicks.

Defenses

Adult dragonflies avoid being eaten by being fast and skilful fliers. They also have excellent eyesight and can see predators coming from any direction.

Dragonfly nymphs protect themselves by hiding, by moving slowly and by being difficult to see.

Damselflies are always watching out for predators.

A dragonfly can see all around, even behind. This makes them very hard to sneak up on.

Weird and wonderful dragonflies

Welcome to the wonderful world of bizarre and extraordinary dragonflies and damselflies!

Prehistoric dragonflies

The largest insect ever to exist was a prehistoric dragonfly. It lived around 250 million years ago and had a wing span of 70 centimeters (27 inches).

Dragonflies have been around for more than 270 million years. They were alive long before the dinosaurs evolved.

The largest dragonfly

One of the largest dragonflies alive today lives in Australia. It has a wing span of 16 centimeters (6 inches).

Distance fliers

Dragonflies can fly long distances. If they find themselves in an unsuitable place they will keep flying. Some strong fliers have even turned up on the decks of ships out at sea — were it not for the ship they would probably have drowned. This shows that they fly in any direction, even out to sea, not really knowing what lies ahead.

Damselflies have wings that beat independently of one another. Not many insects do this.

Surviving drought

Some dragonfly nymphs can survive after their source of water dries up by burying themselves in mud and waiting until the water returns. Some dragonfly eggs can also survive this way. They stay buried and do not hatch until water returns.

Did you know?

Mudeyes are a popular food in some countries.

Tropical nymphs

Some tropical dragonfly nymphs live in water-filled tree hollows or damp rainforest leaf litter.

A sign of a healthy environment

If a freshwater habitat has many different dragonflies and damselflies, then scientists know that it must be a healthy habitat with lots of other animals and plants living there.

Collecting and identifying dragonflies

There is still a lot to learn about the life cycle and behavior of dragonflies.

An example of each kind of dragonfly and damselfly that has been found is kept in collections at museums. These collections are studied by scientists. Sometimes new dragonflies and damselflies are discovered in museum collections long after they were collected because there are so many to examine.

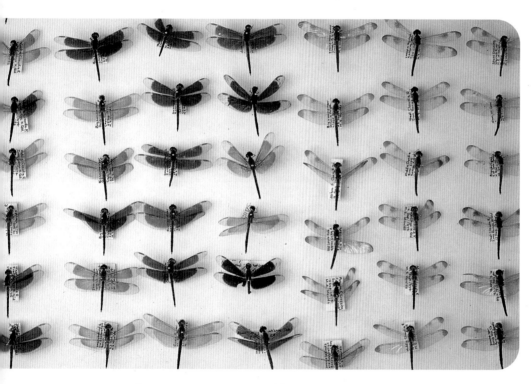

Dragonflies in collections are usually pinned, like butterflies. Sometimes a toothpick is put into the dragonfly's abdomen to keep it straight. Dead insects dry quickly and do not rot. Pinned insects last longer because they do not bump around in containers. The pin gives scientists something to hold when they want to examine the dragonfly.

When scientists collect dragonflies and damselflies, they use special equipment. They use butterfly nets with extra-large hoops to catch them. They use watertight jars and containers to put nymphs in.

Dragonflies are very difficult to catch because they are such fast fliers. Scientists can learn a lot about their behavior just by watching them with a pair of binoculars. They can do this because the dragonflies are large and brightly colored, and the males are usually a different color to the females. This makes them easy to see and means that dragonflies are often studied just like birds.

How are dragonflies and damselflies identified?

Dragonflies are identified by looking very carefully at their shape, size and color. If a dragonfly's shape, size and color is different to all other dragonflies that scientists already know, then this dragonfly is considered a new kind of dragonfly and is given a scientific name.

What do scientists study about dragonflies and damselflies?

After a dragonfly has been given a name, scientists then study:

- where it lives
- what it eats
- how it finds water
- what its predators are
- how its unusual mating behavior works
- what poisons or pollutants kill it or interfere with its normal life cycle.

Did you know?

The pattern on a dragonfly's long, stick-like abdomen is used by scientists to help identify what kind of dragonfly it is.

A scientist studying dragonflies.

Ways to see dragonflies and damselflies

Here are some ways that you can look more closely at dragonflies and damselflies.

Look for dragonflies in your garden. Look for them over a pond, a creek, a dam or a lake.

Look at the different colors in different kinds of dragonflies. Some have red bodies, others have blue bodies or yellow and black striped bodies. Sometimes the wings are colored too.

The best way to get a really good look at dragonflies and damselflies is to find them when they are sitting still. Here are some of the times that you can look closely at dragonflies and damselflies.

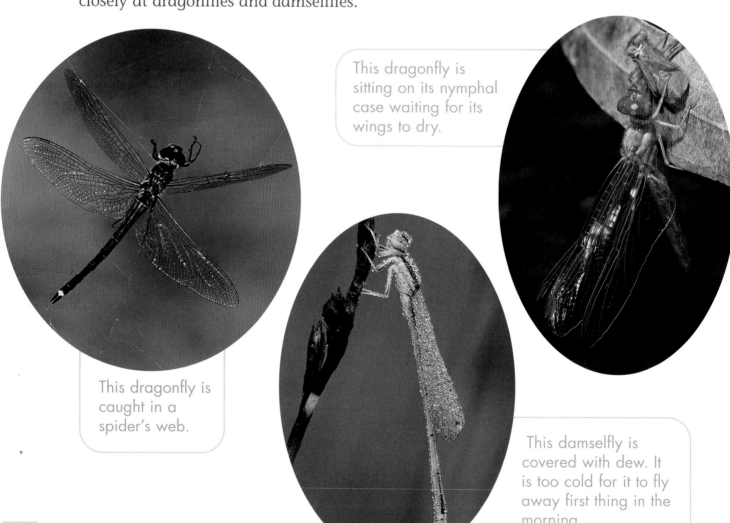

This dragonfly is sitting on its nymphal case waiting for its wings to dry.

This dragonfly is caught in a spider's web.

This damselfly is covered with dew. It is too cold for it to fly away first thing in the morning.

Observing dragonflies

- Find a pond or stream that has dragonflies. Tie a cork to a string on the end of a stick (like a fishing line) and dangle it near a dragonfly.

- Watch the dragonfly. Does it chase the cork thinking that it might be suitable to eat? Or is it a male and does it fly at the cork to defend its territory?

- Watch the dragonfly chase anything that flies, even birds as they fly through.

- Catch a dragonfly. Look at its big eyes and its legs that face forward.

- Listen to the loud noise it makes with its big wings.

Did you know?

Although they may look and sound fearsome, dragonflies and damselflies are completely harmless to humans. They do not sting or bite.

Look closely at the dragonfly's big compound eyes and its cage of legs.

Damselflies can be found around ponds and streams. Look for them sitting on reeds around the water.

Dragonflies and damselflies quiz

1 How many wings do dragonflies have?

2 To which part of a dragonfly's body are its legs attached?

3 What is the special name scientists use for all dragonflies and damselflies?

4 Are the front and back wings of a dragonfly similar or different in shape?

5 Are damselflies strong or weak fliers?

6 How many tails does a damselfly nymph have?

7 Which have the largest eyes, dragonflies or damselflies?

8 Which way do a dragonfly's legs face?

9 Where do dragonflies live?

10 What is a dragonfly's most important sense?

11 What are dragonfly nymphs commonly called?

12 Do dragonflies continue to grow once they have become an adult?

13 Where does a male dragonfly grab a female dragonfly before mating?

14 Where do mudeyes live?

15 Can dragonflies fold their wings back along their bodies?

Check your answers on page 32.

There are no sharp parts on dragonfly bodies that can prick human skin. They are safe to touch.

Glossary

abdomen	The rear section of the body of an animal.
antennae	The two 'feelers' on an insect's head that are used to feel and smell. (Antennae = more than one antenna.)
anus	The hole in the rear of an animal.
caddisfly	An insect like a moth, but with hairs on its wings instead of scales.
damselflies	One of the two kinds of Odonata.
defenses	The ways that animals and plants protect themselves from predators.
habitat	A place where an animal or a plant lives.
larvae	Caterpillars, grubs and maggots are kinds of larvae. In the life cycle of an insect the larval stage is after the egg stage and before the pupal stage. Larvae hatch out of eggs, grow and then turn into pupae. (Larvae = more than one larva.)
mandibles	Hard structures for biting.
membrane	A thin layer of skin.
midge	A small fly.
molting	When an animal sheds its entire skin it molts. The process is called molting.
mudeye	A dragonfly nymph.
nymph	In the life cycle of an insect (like bugs and dragonflies), eggs hatch into little nymphs that grow into adults.
Odonata	The scientific name for dragonflies and damselflies.
predator	An animal that attacks and eats other animals (adjective = predatory).
sexual reproduction	When a male and female living thing combine to make more living things.
sperm	The male reproductive cell.
territorial	When an animal moves backwards and forwards as if patrolling a particular space or when they chase intruders. This is usually called territorial behavior.
thorax	The middle section of an animal's body.

Index